ANIMAL GROUPS

A CARAVAN OF CAMELS

BY SETH LYNCH

Gareth Stevens PUBLISHING

Please visit our website, www.garethstevens.com. For a free color catalog of all our high-quality books, call toll free 1-800-542-2595 or fax 1-877-542-2596.

Library of Congress Cataloging-in-Publication Data
Names: Lynch, Seth, author.
Title: A caravan of camels / Seth Lynch.
Description: New York : Gareth Stevens Publishing, [2022] | Series: Animal groups | Includes index. | Audience: Ages 6 | Audience: Grades K-1
Identifiers: LCCN 2020028902 (print) | LCCN 2020028903 (ebook) | ISBN 9781538263471 (library binding) | ISBN 9781538263457 (paperback) | ISBN 9781538263464 (set) | ISBN 9781538263488 (ebook)
Subjects: LCSH: Camels–Juvenile literature. | Camels–Behavior–Juvenile literature.
Classification: LCC QL737.U54 L96 2022 (print) | LCC QL737.U54 (ebook) | DDC 599.63/6215–dc23
LC record available at https://lccn.loc.gov/2020028902
LC ebook record available at https://lccn.loc.gov/2020028903

Published in 2022 by
Gareth Stevens Publishing
111 East 14th Street, Suite 349
New York, NY 10003

Designer: Tanya Dellaccio
Editor: Kristen Nelson

Photo credits: Cover TARIQ_M_1/RooM/Getty Images; p. 5 Anna Gibiskys/Shutterstock.com; p. 7 Black Kings/Shutterstock.com; p. 9 Dmitry Pichugin/Shutterstock.com; p. 11 HainaultPhoto/Shutterstock.com; p. 13 Alexandra Lande/Shutterstock.com; p. 15 Dr. Juergen Bochynek/Shutterstock.com; p. 17 Shakir Cherooth/Shutterstock.com; p. 19 Frans van Veen/Shutterstock.com; p. 21 oksana.perkins/Shutterstock.com.

Printed in the United States of America

CPSIA compliance information: Batch #CSGS22: For further information contact Gareth Stevens, New York, New York at 1-800-542-2595.

CONTENTS

Boldface words appear in the glossary.

Part of the Pack

People have used camels as pack animals for thousands of years! Most camels today are domesticated, or kept as working animals. Many of these camels live in India or eastern Africa. Others are found in the Middle East, China, and Australia.

Kinds of Camels

There are three kinds of camels. Today, the dromedary, or Arabian, camel is only found as a domesticated animal. It's no longer found in the wild. Most of the world's camels are dromedaries. Dromedary camels only have one **hump**!

There are also domesticated Bactrian camels and wild Bactrian camels. Bactrian camels have two humps! Other than the number of humps, all camels look alike. They are tall and have sandy-colored fur. They also have large eyes and lips.

The Caravan

A herd, or group, of camels is sometimes called a caravan. A caravan is led by one adult male. The rest of the herd is females and young. Some caravans are made up of only male camels. These are called bachelor herds.

Camels are very **social** animals. They **communicate** with each other through many sounds, including groaning and roaring. Camels say hello by blowing other camels in the face! They use their tail, ears, and even head to communicate too.

Camel Calves

The leader of a camel caravan is often the **mate** of the females of his group. Mothers commonly have just one baby, or calf, after about 13 months. Babies don't have humps yet. But they can walk about a half an hour after they're born!

Camel calves drink milk from their mothers for one to two years. They stay with their mothers until they are full-grown, as long as about five years. Then, they're adults. They're able to find mates and have calves of their own.

Time to Eat!

Grown camels eat plants. Their special split lip allows them to eat short grasses and even **thorns**. They are ruminants, which means they bring their food back up after eating it and chew it again!

In Danger

Camels can be seen in zoos throughout the world. But the wild Bactrian camel is **endangered**. There are fewer than 1,000 left on Earth. Some groups are working to save this animal. Still, these wild caravans are in trouble!

GLOSSARY

communicate: to share ideas and feelings through sounds and movement

endangered: in danger of dying out

hump: a part of a camel's body in which it stores fat

mate: one of two animals that come together to make babies

social: able and willing to form groups to live together

thorn: a sharp point found on the side of some plants

FOR MORE INFORMATION

BOOKS

Topacio, Francine. *Creatures in a Hot Desert*. New York, NY: PowerKids Press, 2020.

Vilardi, Debbie. *Why Do Camels Have Humps?* Mendota Heights, MN: Pop!, 2019.

WEBSITES

National Geographic: Arabian Camel

www.nationalgeographic.com/animals/mammals/a/arabian-camel/

Read more about where camels live, what they eat, and how they survive in the deserts where they live.

San Diego Zoo: Camel

animals.sandiegozoo.org/animals/camel

Find out all you want to know about camels here.

Publisher's note to educators and parents: Our editors have carefully reviewed these websites to ensure that they are suitable for students. Many websites change frequently, however, and we cannot guarantee that a site's future contents will continue to meet our high standards of quality and educational value. Be advised that students should be closely supervised whenever they access the internet.

INDEX